Yadira Suárez

Mathematical Innovation: Competencies and Gamification in the Classroom

Yadira Suárez

Mathematical Innovation: Competencies and Gamification in the Classroom

Transforming Mathematical Learning through Gamification:
Strategies for Comprehensive Assessment

ScienciaScripts

Imprint

Any brand names and product names mentioned in this book are subject to trademark, brand or patent protection and are trademarks or registered trademarks of their respective holders. The use of brand names, product names, common names, trade names, product descriptions etc. even without a particular marking in this work is in no way to be construed to mean that such names may be regarded as unrestricted in respect of trademark and brand protection legislation and could thus be used by anyone.

Cover image: www.ingimage.com

This book is a translation from the original published under ISBN 978-613-9-40921-1.

Publisher:
Sciencia Scripts
is a trademark of
Dodo Books Indian Ocean Ltd. and OmniScriptum S.R.L publishing group

120 High Road, East Finchley, London, N2 9ED, United Kingdom
Str. Armeneasca 28/1, office 1, Chisinau MD-2012, Republic of Moldova, Europe
Printed at: see last page
ISBN: 978-620-7-92104-1

Dedication

To my dear students, an inexhaustible source of inspiration and a source of constant learning.

To my teaching colleagues, for their tireless dedication and passion for education, who every day seek new ways to ignite the spark of knowledge in their classrooms.

To my family, for their unconditional support and unwavering love, who have been my pillar and my guide in every step of this journey; especially to my son Dorian and beloved husband Mario.

And finally, to all those who believe in the transformation of education and the power of critical and creative thinking, in the hope that this book will contribute to building a better future for our students.

Table of Contents

MATHEMATICAL INNOVATION: COMPETENCIES AND GAMIFICATION IN THE CLASSROOM

Transforming Mathematical Learning through Gamification: Strategies for Integral Assessment

PRESENTATION

Critical and computational thinking is fundamental in the learning of mathematics, significantly influencing the development of mathematical competencies and student satisfaction. However, the lack of clarity in curricular documents on the implementation of these competencies affects students' development. The STEAM approach, with its inclusiveness and multidimensionality, highlights the importance of integrating logical-formal thinking and mathematics to foster critical thinking and connect students with various academic disciplines.

This book also addresses a study focused on describing how teachers at the Eloy Alfaro Educational Institution use gamification to teach mathematics to children. It has identified the techniques they use and explored the benefits of this strategy in the teaching and learning process. Gamification, by incorporating game elements in education, seeks to make the learning experience more engaging and participatory, significantly improving students' academic performance. This strategy creates a positive dynamic with the subject, awakens interest in mathematics and raises the problem-solving capacity in the context in which students develop.

This book explores the integration of innovative approaches such as STEAM and gamification, highlighting how they can transform the teaching of mathematics. By combining critical and computational thinking with gamified techniques, it seeks to offer students a more complete, meaningful education aligned with the challenges of the 21st century. The studies presented provide detailed insights into current educational practices and offer recommendations for improving mathematics teaching and learning through inclusive and multidisciplinary approaches.

BY THE AUTHOR

Yadira Tatiana Suárez Folleco

Master's Degree in Education and Development Projects with a Gender Approach - Universidad Central del Ecuador

Bachelor of Science in Education with a major in Mathematics and Physics - Universidad Central del Ecuador

Independent researcher

yadira9tatiana@gmail.com

https://orcid.org/0009-0002-0639-2722

Chapter 1: Assessment of Mathematical Learning from a Competency-Based Approach in Secondary Education

Summary

This study addresses the transformation in the teaching and evaluation of mathematics towards a competency-based approach. This change responds to the need for a more comprehensive and applied education, focused on developing skills, knowledge and attitudes that enable students to effectively face the challenges of everyday life and the world of work. Traditional mathematics assessment, dominated by written tests, is complemented by more varied and flexible methods such as performance tests, portfolios, projects, direct observation and rubrics. These tools make it possible to assess not only the final product but also the thinking process and the skills applied by the student. The implementation of competency-based assessment implies a change in the role of the teacher, who must act as a facilitator of learning, and poses challenges in terms of standardization and comparability of results. The research was based on an exhaustive bibliographic review of scientific and academic literature related to the topic, using a qualitative documentary approach. The findings suggest that competency-based assessment in mathematics allows for greater flexibility, adaptability to individual student needs, and a more complete view of learning, promoting more meaningful and lasting learning.

Introduction

Secondary education is a crucial stage in the academic and personal development of students. During this period, the foundations are laid for critical thinking, problem solving and the application of knowledge in real situations. In this context, mathematics plays a fundamental role, not only as a discipline in itself, but also as an essential tool for other areas of knowledge. However, the way in which this subject is taught and assessed has been the subject of debate and reflection in recent decades. The assessment of mathematical learning from the competency-based approach emerges as a response to the need for a more comprehensive and applied education (Castrillo, 2022).

The competency-based approach to education seeks to develop in students a set of skills, knowledge and attitudes that will enable them to effectively face the challenges of daily life and the world of work. In the case of mathematics, this implies going beyond the memorization of formulas and procedures, to focus on the ability to solve problems, reason logically, communicate mathematical ideas and apply concepts in varied contexts (Quiñones Ramírez, Zárate-Ruiz, Miranda-Aburto, & Sosa Celi, 2021).

Traditionally, assessment in mathematics has been dominated by written tests that prioritize the reproduction of procedures and the obtaining of correct results. However, a competency-based assessment requires more varied and flexible methods that allow the assessment not only of the final product, but also of the thinking process and the skills applied by the student.

Among the evaluation strategies most commonly used in this approach are performance tests, portfolios, projects, direct observation and rubrics. These tools allow the teacher to obtain a more complete view of student learning, evaluating aspects such as creativity, reasoning, argumentation skills and the application of knowledge in new situations (Moncini Marrufo & Pirela Espina, 2021).

The implementation of competency-based assessment in mathematics also implies a change in the role of the teacher, who must adopt a more flexible and open attitude, acting as a facilitator of learning rather than as a transmitter of knowledge. This requires continuous training and a commitment to pedagogical innovation.

On the other hand, competency-based assessment poses challenges in terms of standardization and comparability of results. The diversity of assessment methods and tools can make it difficult to standardize grading criteria and the interpretation of results. Therefore, it is necessary to establish clear and consensual reference frameworks to guide assessment and ensure fairness and objectivity (Salcedo Rodríguez & Prez Vázquez, 2020).

In this sense, the main objective of this research is to identify and analyze the strategies, tools and challenges associated with this evaluation modality. It seeks to understand how this approach is being implemented in different educational contexts, what impact it is having on student learning, and how the challenges that arise in its application are being addressed. This review will provide educators, researchers, and policy makers with a solid frame of reference for decision making and continuous improvement of assessment practices in mathematics.

Methodology

The methodology used was based on qualitative documentary research. An exhaustive bibliographic review of the scientific and academic literature related to the topic was carried out, consulting specialized journal articles, books, research reports and educational policy documents (Hernández-Sampieri, Fernández- Collado, & Baptista-Lucio, 2017). Those documents that provided relevant and pertinent information for the study were selected, prioritizing works that specifically addressed the assessment of mathematical learning in the context of secondary education and from a competency-based approach.

Subsequently, a qualitative analysis of the content of the selected documents was carried out, identifying the main themes, concepts and arguments related to competency-based assessment in mathematics. Special attention was paid to the assessment strategies proposed, the challenges mentioned and the results obtained in different educational contexts. From this analysis, the findings were synthesized and connections between the different sources were established, discussing the implications of these findings for educational practice and identifying possible areas for improvement and future lines of research.

Results

Table 1 presents a selection of studies related to the teaching and learning of mathematics in secondary education, focusing on different strategies and tools for the development of mathematical competencies. The following is an analysis of the studies presented:

Table 1: Selection of articles for review

N°	Title	Authors	Link
1.	Flexible teaching and learning of mathematics in rural secondary education.	Vilchez Guizado, J., & Ramón Ortiz, J. Angela (2022)	https://doi.org/10.21556/edutec.2022.80.2431
2.	Inverted classroom: implications in the development of mathematical competencies in secondary education.	Vilchez Guizado, Jesús, & Ramón Ortiz, Julia Ángela (2020).	http://scielo.sld.cu/scielo.php?script=sci_arttext&pid=S 1990-86442020000500225&lng=en&tlng=en.
3.	Quizizz in the development of mathematical competencies in secondary school students: A theoretical review.	Farfán-Pimentel, J. F., Valdez-Asto, J. L., Serveleon-Quincho, F., Asto-Huamaní, A. Y., Carreal-Sosa, C. L., & Farfán-Pimentel, D. E. (2023).	https://doi.org/10.37811/cl_rcm.v7i2.5541
4.	Process of critical and computational thinking in the learning of mathematics.	Ramón Ortiz, J. Angela, & Vilchez Guizado, J. (2023).	https://revistaprismasocial.es/article/vie w/4776

5.	Theoretical foundations for the development of mathematical competencies in Secondary Basic Education.	Gómez-Moreno, F. (2023).	https://doi.org/10.62 697/rmiie.v2i1.27
6.	Logical-Mathematical Thinking review of the STEAM assessment model for developing mathematical competencies	Manuela Angélica Mamani García, Gloria Martínez Gonzales, Jesús María Mamani García de Quedena, Araceli Elizabeth Montero Carcelén (2023).	https://dialnet.unirio j a.es/servlet/article ?code=8827077
7.	Virtual platforms in the development of mathematics competencies in 3rd grade high school students.	Iglesias Albarrán, L. M., Pascual Gómez, I., & Arteaga-Martínez, B. (2020).	https://doi.org/10.55 85/dialogia.n36.182 79

Source: own elaboration

The study "Flexible teaching and learning of mathematics in rural secondary education" by Vilchez Guizado and Ramón Ortiz (2022) focuses on analyzing the implications of flexible mathematics teaching on the learning achievement of fifth grade secondary school students in rural contexts in the province of Huánuco, Peru, during the year 2021, a period marked by the COVID- 19 pandemic. The research adopts a mixed approach, using a non-experimental design and a concurrent triangulation strategy to collect and analyze quantitative and qualitative data.

The results indicate that the majority of students (72%) were satisfied with the flexible teaching received, and more than 67% reached expected and outstanding levels of achievement in the learning of mathematical content taught in a personalized manner. Flexible teaching, which combines face-to-face and virtual modalities (synchronous and asynchronous), was perceived positively by both students and teachers, who highlighted its adaptability to individual needs and its ability to motivate interest and facilitate learning.

The study reveals that flexible teaching is directly related to and positively influences the mathematics learning process in rural secondary school students. The findings suggest that this teaching modality allows greater flexibility in time management, the use of digital resources, and adaptation to the socio-cultural and geographical conditions of the students. In addition, the importance of personalized accompaniment and support by the teacher in the learning process is highlighted.

Another study by Vilchez Guizado and Ramón Ortiz (2020) focuses on analyzing the effectiveness of the inverted classroom didactic model in the teaching-learning process of mathematics in students in the fifth grade of secondary education. The research adopts a mixed approach, using a pre-experimental design to apply the inverted classroom model during the development of mathematics curricular content with activities inside and outside the classroom.

The results obtained with the didactic model confirm its effectiveness for learning mathematics, with more than 65% of the students achieving between excellent and good results according to the evaluation rubric. In addition, more than 70% of the participants showed full satisfaction with the didactic strategy. The mathematical competence of the students through the inverted classroom is considered to be at a superlative level, and the level of satisfaction with their learning achieved and the mathematical competencies developed was rewarding.

The inverted classroom is characterized by a reorganization of learning times and spaces, where the time dedicated to the expository class is moved out of the classroom, and class time is used to carry out activities that require more active participation on the part of the student. This pedagogical model seeks to facilitate student participation in activity-based learning and to encourage exploration, articulation and application of ideas during class time.

In the same vein, the study by Farfán-Pimentel et al. (2023) addresses the relevance of incorporating technological tools in the teaching-learning process to make it more dynamic, creative and innovative. In particular, it focuses on the Quizizz application, a tool that allows students to learn in a playful and motivating way, developing mathematical abilities and skills essential for everyday life.
The research is theoretical in nature and is based on the method of analysis and synthesis, using an exhaustive search of materials such as thesis reports, scientific articles, specialized literature, databases and information platforms related to the topic of study. The main objective is to analyze the Quizizz strategy in the development of mathematical competencies in high school students.

The results of the literature review indicate that the use of Quizizz as a pedagogical tool in virtual environments facilitates the development of mathematical competencies in elementary and high school students. It is highlighted that this tool promotes a culture of constant evaluation and allows formative feedback from the teacher, generating in the student a process of continuous improvement.

The study "Process of critical and computational thinking in the learning of mathematics in secondary education" by Ramón Ortiz and Vilchez Guizado (2023) focuses on analyzing the incidence of critical and computational thinking in the learning of mathematical concepts and procedures in secondary education students. The research adopts a mixed approach, combining qualitative and quantitative methods, and uses a non-experimental design. The data collection instruments include a Likert-type questionnaire and a cognitive test, and data analysis is performed using descriptive statistics.

The results of the study show a sustained development of critical and computational thinking in students during mathematics learning activities, expressed in the fluency of handling concepts and logical procedures in problem solving. Statistically, a correlation was found between the level of learning achieved and the development of critical and computational thinking, with values of 0.545 and 0.823 respectively.

The study concludes that the computational and critical thinking developed by secondary education students significantly influences learning and the development of mathematical competencies, with implications in the level of satisfaction with their academic and personal achievements. This research highlights the importance of integrating critical and computational thinking in the teaching-learning process of mathematics in secondary education to improve learning and the development of mathematical competencies in students.

The article by Fabio Gómez-Moreno (2023) presents an exploration of the theoretical principles that support the teaching and learning process of mathematics in Secondary Basic Education, particularly in Colombia. Mathematical competencies and how they are developed in this educational process are analyzed, based on relevant theories and Colombian educational regulations.

The author stresses the importance of relating mathematical learning contents with the daily life of students through problematic situations, although he points out that the Colombian Curricular Guidelines do not provide clear guidelines to stimulate students' interests towards the plurality of knowledge and its link with the productive sector. This, according to Gómez-Moreno, does not favor human, social and technological development, nor does it contribute to the inclusion of students in the resolution of problems in their context.

The Basic Competency Standards, built from the Curricular Guidelines, establish the parameters of what every student should know and do to achieve an expected level of quality in their mathematical education. However, the author criticizes the fact that these competencies are not clearly stated in the document, which hinders their implementation in curricular planning.

Gómez-Moreno also examines other guiding documents such as the Basic Learning Rights, the Reference Matrix and the Learning Grids, highlighting that, although these are oriented to the development of mathematical competencies, they focus on mathematical content without sufficiently addressing attitudinal aspects or "knowing how to be".

The author concludes that, although the Colombian educational system is geared towards the development of mathematical competencies, the curricular documents do not offer a clear guide on how to achieve this objective, significantly affecting the development of mathematical competencies in students in Secondary Basic Education.
Likewise, the study "Logical-Mathematical Thinking: Review of the STEAM assessment model to develop mathematical competencies" by Mamani García et al. (2023) examines the development of mathematical competencies in high school students using an assessment model based on the STEAM (Science, Technology, Engineering, Art and Mathematics) approach. The research highlights the importance of integrating logical-formal thinking and mathematics to foster critical thinking and connect students to various academic disciplines.

The proposed assessment model is multidimensional stochastic and inclusive in nature, encompassing aspects such as gamification, project-based learning, problem-based learning and didactic engineering. These pedagogical strategies are supported by sociocultural and dialogic learning theories, with a formative assessment approach that emphasizes assessment for the achievement of students' potentialities.

The study concludes that the computational and critical thinking developed by secondary education students significantly influences learning and the development of mathematical competencies, with implications on the level of satisfaction with their academic and personal achievements. The research highlights the importance of integrating critical and computational thinking in the teaching-learning process of mathematics in secondary education to improve learning and the development of mathematical competencies in students.

On the other hand, the research by Iglesias Albarrán, Pascual Gómez and Arteaga-Martínez (2020) analyzes the adaptation to fully digital learning in a secondary school in southern Spain during the COVID-19 pandemic confinement. It focuses on algebra learning in 14-15 year-old students, using metacognitive strategies and digital materials to foster autonomy and teacher-student communication. Results indicate that students exceeded the evaluation criteria and that the design facilitated optimal feedback during the teaching-learning process. The study highlights the importance of adapting teaching to digital contexts and using metacognitive strategies to improve algebra learning in secondary education.

Conclusions

Research on the assessment of mathematical learning from the competency-based approach in secondary education reveals that adaptability and personalization of instruction are crucial to meet the individual needs of students. Innovative pedagogical models such as the flipped classroom have proven to be effective in improving mathematics learning, encouraging participation and the development of high-level competencies. The integration of technology, especially tools such as Quizizz, facilitates the development of mathematical competencies in virtual environments, promoting a culture of constant evaluation and formative feedback.

Critical and computational thinking is fundamental in the learning of mathematics, significantly influencing the development of mathematical competencies and student satisfaction. However, the lack of clarity in curricular documents on the implementation of mathematical competencies affects the development of these competencies in students. The STEAM approach, with its inclusiveness and multidimensionality, highlights the importance of integrating logical-formal thinking and mathematics to foster critical thinking and connect students with diverse academic disciplines.

Adaptation to fully digital learning environments is possible and can facilitate optimal feedback and achievement of assessment criteria, especially in situations of confinement such as the COVID-19 pandemic. In conclusion, the assessment of mathematical learning from the competency-based approach promotes more meaningful and lasting learning, adapting to the specific needs and contexts of students. It is necessary to continue researching and developing pedagogical and assessment strategies that promote the comprehensive development of mathematical competencies in students.

Bibliographic references

Castrillo, C. J. H. (2022). Methodologies for competency-based learning of Applied Differential Equations in Physics by using technology in the Mathematical Physics career. Revista Torreón Universitario, 11(32).

Farfán-Pimentel, J. F., Valdez-Asto, J. L., Serveleon-Quincho, F., Asto-Huamaní, A. Y., Carreal-Sosa, C. L., & Farfán-Pimentel, D. E. (2023). Quizizz in the development of mathematical competencies in high school students: A theoretical review . *Ciencia Latina Revista Científica Multidisciplinar, 7*(2), 2987-3005.
https: //doi.org/10.37811/cl rcm.v7i2.5541

García, M. A. M. M., Gonzales, G. M. M., de Quedena, J. M. M. M. G., & Carcelén, A. E. M. (2023). Logical-mathematical thinking: review of the STEAM assessment model to develop mathematical competencies. Revista de filosofía, 40(103), 83-98.

Gómez-Moreno, F. (2023). Theoretical foundations of the development of mathematical competencies in Secondary Basic Education. *Revista Mexicana De Investigación E Intervención Educativa, 2*(1), 5-15. https://doi.org/10.62697/rmiie.v2i1.27

Hernández-Sampieri, R., Fernández-Collado, C., & Baptista-Lucio, P. (2017).
Differences between quantitative and qualitative approaches.

Iglesias Albarrán, L. M., Pascual Gómez, I., & Arteaga-Martínez, B. (2020). Learning algebra in Secondary Education: metacognitive strategies from digital technology. *Dialogia,* (36), 49-72.
https://doi.org/10.5585/dialogia.n36.18279 127

Moncini Marrufo, R., & Pirela Espina, W. (2021). Virtual teaching strategies used with higher education students for meaningful learning. SUMMA, 3(1), 1-28. https://doi.org/10.47666/summa.3.1.13

Quiñones Ramírez, Leonela, Zárate - Ruiz, Gustavo, Miranda - Aburto, Elder, & Sosa Celi, Paul (2021). Competency Approach (CE) and Formative Assessment (EF). Case: Rural school. *Propósitos y Representaciones, 9*(1), e1036.
https://dx.doi.org/10.20511/pyr2021.v9n1.1036

Ramón Ortiz, J. Ángela, & Vilchez Guizado, J. (2023). Process of critical and computational thinking in the learning of mathematics in secondary education. *Revista Prisma Social,* (41), 194-211. Retrieved from
https://revistaprismasocial.es/article/view/4776

Salcedo Rodríguez, Medalit Nieves, & Prez Vázquez, Mateo Dolores (2020). Relationship between emotional intelligence and mathematical skills in high school students. Mendive. Journal of Education, 18(3), 618-628. Epub September 02, 2020. Retrieved March 28, 2024, from.
http://scielo.sld.cu/scielo.php?script=sci arttext&pid=S 1815-76962020000300618&lng=es&tlng=es.

Vilchez Guizado, J., & Ramón Ortiz, J. Ángela (2022). Flexible teaching and learning of mathematics in rural secondary education. *Edutec. Electronic Journal Of Educational Technology,* (80).
https://doi.org/10.21556/edutec.2022.80.2431

Vilchez Guizado, Jesús, & Ramón Ortiz, Julia Ángela (2020). Inverted classroom: implications in the development of mathematical competencies in secondary education. *Conrado, 16(76),* 225-233. Epub 02 October 2020. Retrieved on March 28, 2024, from http://scielo.sld.cu/scielo.php?script=sci arttext&pid=SiQQo-86442020000500225&lng=es&tlng=es.

Chapter 2: Use of Gamification in the Teaching of Mathematics by teachers of the Eloy Alfaro Educational Institution in the third quarter of the school year 2023 - 2024.

Summary

The research focused on the use of gamification as an effective strategy in the teaching of mathematics for teachers at the Eloy Alfaro institution. The objectives included describing the use of gamified strategies, identifying specific techniques and analyzing the benefits of gamification. The results highlighted a high consistency in the instrument used, which included surveys applied and analyzed using SPSS statistical software. It was observed that a significant percentage of teachers related learning experiences to gamification, reflecting contextualized planning in the educational environment and student-centered attention. After designing gamified systems, teachers reported an increase in academic achievement in Mathematics. The frequency of use of gamified experiences in the classroom was notable, with 39.2% of teachers reporting an improvement in their students' problem-solving competence at the end of the strategy. It was concluded that gamification is an essential dynamic for learning in the mathematics classroom.

Introduction

Today, education emerges from an increasingly dynamic world with new technologies, which offer unique opportunities to transform traditional teaching and learning methods. Gamification as an innovative strategy seeks to involve students actively and consciously in their learning process, especially in the subject of Mathematics.

This study focuses on describing the use of Gamification in Mathematics Teaching by teachers of the Eloy Alfaro Educational Institution in the third quarter of the school year 2023 - 2024. This institution seeks to improve educational quality and the incorporation of gamification is the first step for this progress; it is supported by studies:

The research work carried out by Culqui (2023) with the objective of designing a methodological strategy through gamification to strengthen the academic reinforcement process in the subject of Mathematics for children from 5 to 9 years of age. Descriptive research, with a qualitative-quantitative approach, where bibliographic and field research techniques were applied with the use of an observation sheet and a survey to 30 students of the educational center.

The gamification proposal of León (2022) in a didactic unit in the subject of mathematics with students of 6th grade EGB. Research focused on designing a gamification proposal for the area of mathematics for students in the 6th year of EGB, conducting a workshop consisting of 7 sessions of 5 hours each, in 7 weeks. With the objective of achieving growth, teacher updating and the satisfaction of those involved in the integrative project; for this purpose, an intervention proposal with a final evaluation was carried out.

The study conducted in Santo Domingo, Ecuador by Vásquez (2021), a quantitative approach research, applied, correlational and non-experimental cross-sectional design, used the deductive method by applying questionnaires, one for the gamification variable and another for the learning standards variable; testing the hypothesis of the study, which states that gamification has a positive influence on the learning of mathematics of 8th, 9th and 10th grade students.

The research of Transversal competencies in the university educational context: a critical thinking from the principles of gamification, written by Polo, Ramirez, Hinojosa and Castañeda (2022) this pre-experimental, descriptive and correlational study, with characterization of the subject; has the purpose of implementing gamification as a strategy that allows the incorporation of new educational practices in the classroom to increase the motivation and commitment of students, to raise critical thinking as an educational innovation of university students through problem-based learning.

Through this research work it is expected to describe the use of gamification in the teaching of mathematics by teachers of the educational institution, identify the gamification techniques used by teachers in the teaching of mathematics and analyze the benefits of the application of gamification in the teaching-learning process of the subject; promoting the effective use of this strategy as an educational resource.

The teaching of mathematics faces great barriers, such as the low level of motivation of students, the lack of interest in learning the subject and the study habits of each student. Gamification applies elements of games in the educational context, adapts them and offers a way to facilitate learning experiences, more attractive, interactive, participatory, collaborative and enriching; through competitions, challenges, rewards and games. Gamification contributes to the academic, formative and personal performance of students.

1.1. Gamification

From several concepts of gamification, it is pertinent to take the definition of García, F., Cara, J.F., Martínez, J.A., and Cara, M.M., (2020) who conclude that "this methodology predisposes students to participate, encouraging their skills and competencies. It is a very powerful tool that completely changes the traditional school perspective and redefines the educational process. From its implementation, the process focuses on the needs of its consumers, in this case the students" (p. 18). In its absence, it can be stated that gamification in the context of teaching is a tool that is used to complement and improve outdated teaching methods from years ago.

1.1.1. Gamification as a teaching-learning technique

López (2019) points out "by way of conclusion it is essential to understand the important role of the game and gamification as a teaching and learning technique and as a potentiator of skills necessary for personal, professional and integral development of the human being and to promote the game from an early age in childhood and throughout the development and growth of the human being. While it is true that the needs of each individual are different, it is also true that, regardless of the age of the human being, this requires the game, being a necessity and a right from the moment he/she is born, becoming a tool to facilitate the acquisition of meaningful knowledge and understanding of the environment" (p. 9).

From the above it can be described that gamification is a technique that uses game elements, not playful in some environments and that is why the activities are more attractive and fun, are stimulating for children at an early age and make their school performance is favorable, promoting autonomous learning, critical thinking, problem solving and teamwork; encourages active study, instant answers, increases security, confidence and strengthens their personality.

1.1.2. Gamification and teaching pedagogy

Teaching performance plays a major role in the educational process. In this regard, Hernández-Peñaranda, J. O., Jaramillo-Benitez, J., & Rincón-Leal, J.

F. (2020) "teachers of the 21st century must use the pedagogical power of gamification which is undoubtedly a valuable resource to seduce the student in the learning of mathematics and at the same time accept the challenge of educational innovation, taking into account that gamification requires creating a narrative that guides the objective of the teaching-learning process that is desired in the classroom" (p. 33).

The commitment of the teacher is to be clear about what he/she wants to achieve by integrating gamification as a complement to pedagogy in teaching, to define learning objectives, a variety of appropriate gamification elements, to design planning with the possibility of evaluating these adaptations and, above all, to be careful not to overuse this resource.

1.2. Gamification Elements

They are generally represented in pyramidal form, according to Chaves Yuste (2019) "dynamics are elements that are present in almost all games and involve the highest level of abstraction, by mechanics are understood more specific elements that involve detailed actions and direct the players towards the desired direction to meet the established objectives and on the other hand the components are elements necessary for the functioning of the game mechanics" (p. 2). These elements make the teaching-learning process more attractive and effective.

Dynamic elements such as challenges, competitions, collaborations, academic reinforcements, feedbacks that motivate to improve school performance can be proposed; on the other hand, it is appropriate to adapt mechanical elements in the evaluations of school activities such as points for completing tasks or participating in class, badges to recognize achievements or important progress, rankings to compare the progress of students and rewards for achieving goals. Gamification components refer to the aesthetics of the games, i.e. design, interface, images, sound... etc. Those that make the experience more engaging and relevant.

1.3. Gamification Features

During the last years of teaching practice in the classroom, it is possible to describe the multiple errors on the part of the teachers, at the beginning the lack of experience makes them more common, then with practice it is possible to understand that one of the shortcomings is the decrease of the stimulus in the first years of learning, where creativity, the development of fine motor skills, relaxation exercises, basic psychomotor exercises for writing, intrinsic motivation and meaningful learning for concentration in areas of logic and mathematics took precedence; then something changes in the teaching process of children in the early school years and their order of advancement, they stop playing learning and there is this divorce on the way, as well as Ardila-Muñoz, J. Y (2019). He points out as characteristics:

Gamification adapts elements of game design to be implemented in non-game situations such as education. In this way, it seeks to create fun and voluntary learning environments. Gamification is based on influencing people's behavior or attitude; in the case of education, it seeks to increase students' commitment to learning. To achieve this, it uses the definition of rules, follow-up tasks and positive feedback (p. 79).

1.4. Benefits of Gamification

A rdila-Muñoz, J. Y. (2019). Mentions "gamification in education brings benefits such as greater control and monitoring of the actions that students advance; evaluative activities lose their punitive nature; the teaching-learning relationship is characterized by competitiveness and cooperation, and promotes problem-based learning and learning by discovery" (p. 79). The implementation of gamification in a correct, designed and detailed way will allow children through the game to generate skills to solve real-life problems with greater skill and judgment.

1.5. Gamification and Mathematics

Through their study, Aristizábal, Colorado & Gutiérrez (2016) recommend "taking into account the educational reality, it is recommended that teachers propose and adopt innovative pedagogical and didactic strategies within the framework of the game as a teaching strategy, which lead to the development of mathematical thinking. It is suggested to give continuity to the game proposal as a didactic strategy to develop numerical thinking in the four basic operations and in other subjects, as an effective strategy to overcome the difficulties encountered in mathematics education. It is suggested to teachers of mathematics in basic education, the application of strategies aimed at developing mathematical thinking in students, to enhance the skills that allow them to improve access to knowledge" (p. 124-125).

1.5.1. Tools for Gamification in Mathematics

There is a great variety of gamification tools used in the teaching of mathematics, of which we point out these:

1.5.1.1. Kahoot kids. - is a gamified learning platform for children ages 4 and up, offering a wide variety of educational activities covering Math, Language Arts, Natural Science and Social Studies, helping children learn and have fun at the same time.

Figure 1: *Platform logo*

Kahoot!

Note. Kahoot platform. Taken from Wikipedia

1.5.1.2. Quizizz. - is a platform that includes gamification elements, such as points, badges and rankings, to motivate students and encourage active learning.

Figure 2: *Platform logo*

QUIZIZZ

Note. Quizizz platform. Retrieved from URL: https://quizizz.com/

1.5.1.3. Educaplay. - Educaplay is a valuable tool for educators who wish to create engaging and interactive learning experiences for their students.

Figure 3: *Platform logo*

Note. educaplay platform. Retrieved from URL: https://es.educaplay.com/

1.5.1.4. Cerebriti. - an online platform that allows users to create and share educational games, is a great tool for teachers and students.

Figure 4: *Platform logo*

Note. cerebriti platform. Retrieved from URL: https://www.cerebriti.com/

1.5.1.5. Matific. - is a digital learning platform designed to help children ages 4-12 learn math in a fun and engaging way. It offers a variety of interactive games, activities and worksheets aligned to curricula.

Figure 5: *Platform logo*

Note. Matific platform. Retrieved from URL: https://www.matific.com/

1.5.1.6. Phet. - Phet interactive simulations are a valuable resource for students of all ages seeking to explore STEM concepts for science, technology, engineering and mathematics education. They are exploratory explanations that allow for interaction.

Figure 6: *Platform logo*

Note: PHET Platform. Retrieved from URL: https://phet.colorado.edu/es/

1.5.1.7. Nearpod. - is an educational technology platform that allows teachers to create interactive presentations for students, with a wide variety of elements such as videos, quizzes, polls, images, virtual reality experiences, etc.

Figure 7: *Platform logo*

Note. cerebriti platform. Retrieved from URL: https://nearpod.com

Design and Method

2.1. Formal object

The objective of this study, is to describe the use of Gamification in the Teaching of Mathematics by the teachers of the Eloy Alfaro Educational Institution in the third quarter of the school year 2023 - 2024 and its influence on the school performance of their students, according to Contreras and Eguia (2017), concludes that "using gamification in the classroom is effective as long as it is used to encourage students to progress through the learning content, to influence their behavior or actions and to generate motivation. It is possible to motivate learners with the introduction of a methodology that includes challenges, goals, etc. These elements encourage participation or action in human beings in general. However, even the cultural context or previous experiences must be taken into account" (p. 16).

Likewise, the advantages of gamification within the teaching-learning process could be mentioned, every teacher should take into account the development process of the gamified subject and its sustainability over time, if the relevance of a start is maintained until the end of the planning will mean that the student's motivation was high and it was a success, hence the following should be to include more skills of this type to strengthen the knowledge of the students and maintain the improvement in their academic performance.

2.2. Type of research

Due to the characteristics of the research, it is located at the descriptive level, a study of the prevalence of gamification in the teaching of the subject of Mathematics, the approach is quantitative due to the analysis of its only variable "gamification in education", for which a measurement was used; that is to say, a transversal treatment of the information obtained was carried out. For data collection, a retrospective search for information from secondary sources was carried out, documenting in local and non-local scientific articles, the authors' observation of the performance and teaching of the institution's teachers, gathering impressions of the work in the third part of the school year. The main function of these studies is to specify the properties, characteristics, profiles of groups, communities, objects or any phenomenon. Data of the study variable are collected and measured (Hernández- Sampieri and Mendoza, 2018). This type of study is not dedicated to the manipulation of the variable, but it is observed, described and substantiated, being able to obtain several results.

2.3. Research design

At the descriptive level, this study is non-experimental, because it determines the characteristics and important elements of the actors in the process, in this case the teachers and based on this the methodology applied to their teaching performance in the teaching of the subject of Mathematics with the gamification strategy, in turn it can be said that it is a correlational research because it seeks to analyze the benefits of the application of gamification to school planning and its impact on academic performance.

2.4. Scope of study

The population consisted of 28 people, including teachers, administrative and service personnel of the Eloy Alfaro educational institution. A sample was not necessary because the entire population was studied, i.e. the 28 teachers of the institution, who teach the subject of mathematics at the high school, elementary and middle school levels. The data collection technique in the research was a survey, which consisted of a questionnaire of 23 questions that measure the variable and its dimensions such as: gamification in teaching, academic performance, academic performance, frequency of use, tools used, competencies or skills developed and didactic strategy linked. The design of the survey instrument was validated with 3 experts, the questionnaire is divided into 8 sections, section A contains socio-demographic data and sections B, C, D, E, F, G and H contain the aforementioned dimensions. The instrument used in this study was made in a form in Google Forms in order to obtain information on the treatment of the variable gamification in the teaching of mathematics.

Table 2: Questions of the Questionnaire

B. Gamification in the teaching	B1. Have you heard what gamification refers to?
	B2. Do you use any technological tool to teach the contents in the classroom?
	B3. Have you used any technological tool based on gamification?
	B4. Are you familiar with digital educational platforms for teaching mathematics?
	B5. Do you plan your classes using gamification for teaching mathematics?
	B6. Do you find the application of gamification in the teaching of mathematics beneficial?
C. Academic Performance	C1. Do you consider that since you have applied the gamification strategy, you have increased the understanding of the concepts explained in the classes?
	C2. Do you consider that since you have applied the gamification strategy, the academic performance in Mathematics has increased?
	C3. Do you prefer to work in mathematics classes using gamification tools?
	C4. Do you prefer to work on academic reinforcement activities in Mathematics using gamification tools?

D. Teaching performance	D1. Do you have in-depth knowledge of the subject you teach?
	D2. Is a teacher a planner, organized, and does he/she assess his/her students effectively?
	D3. Are you a teacher who communicates clearly, accurately, and effectively with students, families, and other educational professionals?
	D4. Are you a teacher with a positive attitude toward teaching and learning for your students?
E. Frequency of use	E1. How often do you use gamification to teach your classes?
F. Tools used	F1. Have you used any of these digital platforms? Kahoot, Quizizz, Educaplay, Cerebriti, Matific, Phet or Nearpod.
	F2. Do you consider that carrying out academic reinforcement activities with the use of cell phones, tablets and computers facilitates learning?
	F3. Do you consider that gamification tools such as: avatars, badges, missions, unlocking of
	content, rewards, leaderboard and levels, are they valuable elements for teaching and learning?
G. Competencies or skills developed	G1. With gamification, which mathematical operations do you master best and which were difficult to teach before?
	G2. In which of the following methodological strategies applied in the area of Mathematics did you use gamification?
	G3. Which competencies do you think were strengthened after using gamification in teaching

	your students?
H. Linked didactic strategy	H1. The learning activities used in gamification were focused on these aspects: game-based learning, challenge-based learning, narrative, immersive, peer-to-peer learning, etc.?
	H2. In the classes, does the teacher use dynamics related to games to motivate learning?

Source: own elaboration

According to Hurtado (2000), one of the most relevant characteristics of the questionnaire is that the questions are asked succinctly and do not require the presence of the researcher or the person who applies it. It is important that the questionnaire is not so extensive, otherwise the respondents may get results that differ from reality. In addition, the questions should be formulated in a simple manner that allows the respondent to answer them in the shortest possible time. This instrument must meet the requirements of validity and reliability before being applied; that is why, once validated by the experts, the pilot test was applied and a Cronbach's alpha of 0.915 was obtained with n=19 using SPSS software, with a high consistency.

Once the instrument was applied to the population, other data collection techniques were used, such as: observation as one of the most effective processes for quantitative research, documentation of reliable bibliographic sources and data analysis by means of statistical programs.

Figure 8:

Estadísticas de fiabilidad

Alfa de Cronbach	N de elementos
,915	19

Fuente: software SPSS

Fieldwork and Data Analysis

The work consisted of applying the instrument designed once it had been validated and the adjustments suggested by the experts had been made, then the information obtained from the Google Forms questionnaire was processed with the help of the Excel statistician, then the information was analyzed and the results were organized; all this took approximately 2 months. For data collection, the survey technique was applied with a Likert-type measurement scale, with a population of 28 teachers of the Eloy Alfaro Educational Institution, in which the dimensions of the single variable, gamification in the teaching of the subject of Mathematics, were measured.

To analyze the data, SPSS version 26.0 statistical software was used, analyzing the data matrix of the pilot test, understanding that the behavior of the results corresponds to the correlation of the items and the high consistency of Cronbach's alpha; that is to say that the study goes hand in hand with the objectives defined in the research.

Quantitative analysis

The results obtained from the survey reveal that 100% of the respondents, teachers of the Eloy Alfaro Educational Institution, answered the questionnaire. Among the questions on associated factors, we can highlight that the average age of a teacher ranges between 37 years, 7.1% corresponds to the percentage of male teachers and 92.9% to the percentage of female teachers, the academic level of teachers is an important element of analysis, since 67.9% of teachers have higher education, 28.6% have postgraduate studies and 3.5% correspond to a teacher with secondary education, on the other hand, the salaries received are level, ranging from $600 to $1000; It should be noted that all teachers have access to the Internet and electronic devices for teaching.

Technology and electronic devices are not necessary to carry out gamification experiences in teaching; gamification is currently being assimilated as game strategies that should be incorporated into teaching planning and should serve to strengthen the practical and functional learning of students, especially at early ages. Figure 9 shows the three main areas in the application of gamification for the subject of Mathematics and it is important to point out that there is a considerable percentage of teachers who relate learning experiences to games. In order to start gamifying, a series of knowledge is necessary to plan the approach to the gamification experience, the first is to create the environment.

Figure 9:

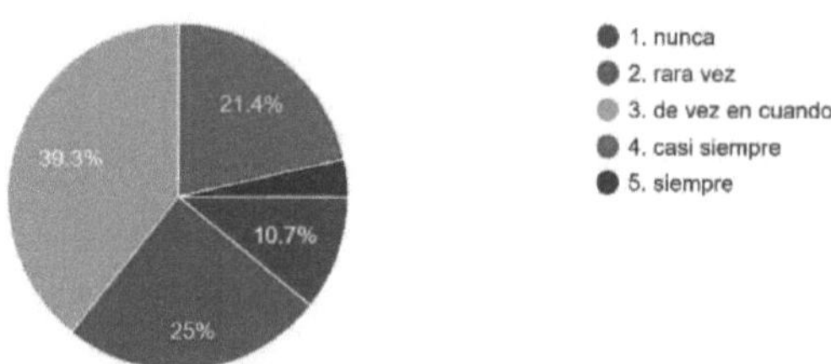

Source: Survey to teachers of the Eloy Alfaro Educational Institution.

Figure 10 represents the data on the teachers' opinions on the benefits of using gamification in the teaching of mathematics and it can be pointed out that with contextualized planning in the educational environment, focusing attention on the students, creating an effective channel of communication between the teacher who delivers the information and the recipients of this gamified context, satisfactory results are obtained that benefit the process, not only cognitively but also the student's human development, creating confidence and security in their personal development.

Figure 10:

Source: Survey to teachers of the Eloy Alfaro Educational Institution.

The results are achieved from experience by selecting the right elements to gamify in the subject, not everything can be gamified, in this case it is necessary to mention the evaluation system for students, who in this scenario are children of early ages, up to 11 and 12 years old, where the basis of the training process is more qualitative than quantitative, each teacher after designing a system, obtains an academic performance in Mathematics that is increasing, thanks to the gamified strategies used for evaluation, which is shown in Figure 11.

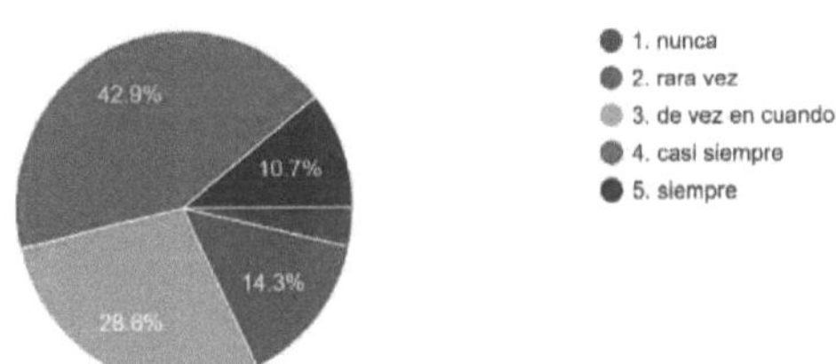

Source: Survey to teachers of the Eloy Alfaro Educational Institution.

The impact of gamification strategies is of relevance in the process of the subject and the most important limitation in this process is to sustain gamification over time, hand in hand with motivation, it is crucial to make an analysis at the beginning, on the way and at the end to determine the degree of interest of the students; a guarantee of this is the frequency with which they work using gamified experiences in the classroom, in Figure 12 it is evident that the teachers of this institution do it every week and it is a result that makes a significant difference for this study.

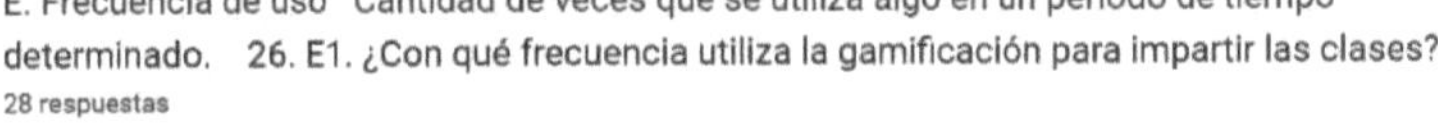
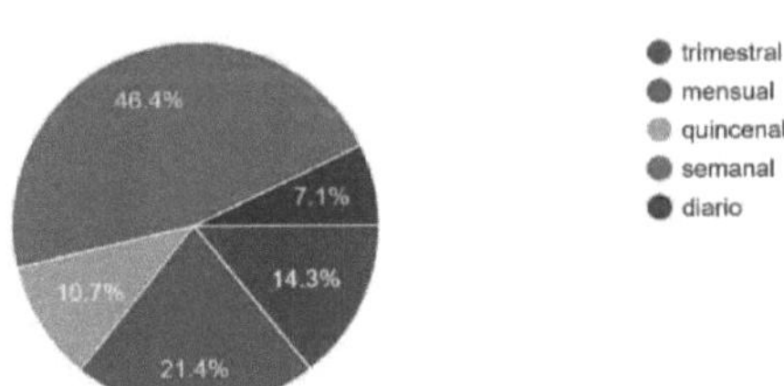

Source: Survey to teachers of the Eloy Alfaro Educational Institution.

One of the questions in the questionnaire referred to the competencies strengthened after using gamification, although there are varied answers in Figure 13, it can be seen that 39.2% of the teachers agree that at the end of the application of the strategy, their students have high problem-solving skills, this reflects that there is mastery of knowledge, it helped to build it in an unconventional way, learning was facilitated, more didactic in the process, it was interactive and aroused interest and above all the objective was reached, which was to better understand the world around them.

Figure 13:

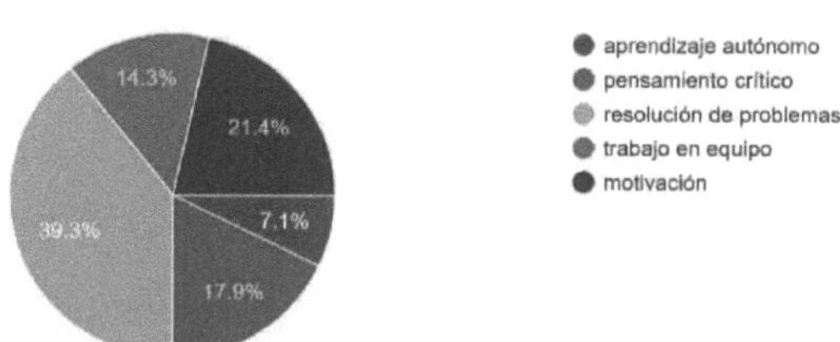

Source: Survey to teachers of the Eloy Alfaro Educational Institution.

Among the responses in Figure 14, on the dynamics related to games to motivate learning, the response with 42.9% corresponding to the scale "always" is notable; The percentages that follow in value of the scales "almost always" and "from time to time" are also relevant, indicating that to a lesser extent there is an interest on the part of the teacher in involving the game in the classroom, but it is still important.

Figure 14:

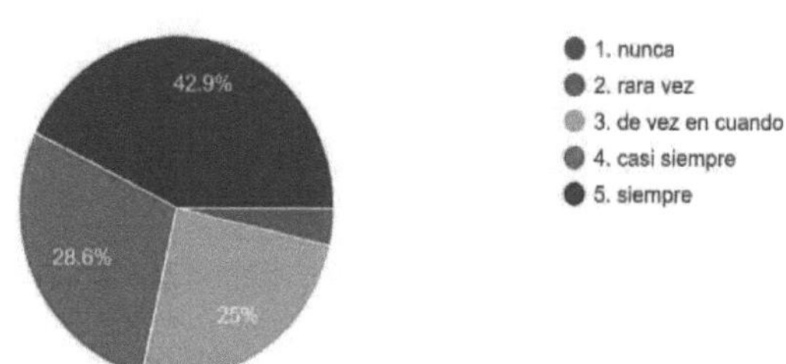

Source: Survey to teachers of the Eloy Alfaro Educational Institution.

Qualitative analysis

Regarding the influence of the use of gamification in the teaching of mathematics by the teachers of the Eloy Alfaro Educational Institution, it can be concluded:

- Teachers design planning with academic rigor that involves play in their learning experiences.

- The gamification strategy allows for greater comprehension by the students.

- Teachers have better evaluation results through gamification.

- Students have an improvement in academic performance and in the feedback process.

- To best reap the benefits of gamification, more frequent use is required.

- It is necessary to overcome the competencies entrenched in the teaching of subjects such as mathematics, so that students not only solve problems, but can create autonomous learning, develop critical thinking, work in teams and maintain their level of motivation.

- From the academic point of view, the teacher must help guide the basic principles in his/her students and this is achieved by using learning dynamics related to the game that allow a better interrelation between the teacher and the student.

Discussion

In recent years, Gamification has been used in multiple educational initiatives to teach and learn Mathematics, proving to be an effective strategy for students, far from being boring, it creates work habits and effort, involves the actors and promotes autonomy for proper problem solving. This study has as a line of research, the study of Gamification in the teaching of Mathematics; it can be mentioned that the strengths of it are the results obtained in the application of its instrument, which allowed us to describe briefly the use of Gamification in the teaching of Mathematics by teachers of the Eloy Alfaro Educational Institution, which largely handle gamified strategies for their learning experiences.

On the other hand, the gamification techniques used by teachers to teach were identified, although it is important to point out that for a future research it would be good to analyze the impact of each one of them in the teaching and learning process of the children involved, with the objective of obtaining a broader perspective of the benefits of the application of gamification, comparing the results obtained from this research and that of Culqui, Daniela (2023); it can be pointed out that gamified activities are a viable alternative for the academic reinforcement process and that it is favorable for children between 5 and 9 years old.

For the teachers who contributed to the research it is important to specify that they should invest more space in their planning to carry out a teaching update, research and design of their classes, from the results it can be concluded that the academic level is higher, but only a low percentage of teachers who have a postgraduate academic level, this will help to give more weight to their teaching work and achieve better conclusions. In her work León, Daniela (2022) studies the growth of teachers and their job satisfaction by making an intervention proposal that improved the level with the results of the final evaluation after Gamification in a didactic unit in Mathematics.

In the study conducted by Ramírez María and Olmos Héctor (2020), within the curriculum of the first grades of study, from elementary to high school, the subject of mathematics is included because of its relevance in the life of every person, both for the calculations and accounts per se, and for its importance in the formation of a brain structure suitable for logical reasoning that provides the ability to solve everyday problems, and given that this has not been the most liked or popular subject, schools and institutions must move from a monotonous and boring teaching-learning of mathematics to a scheme that motivates students and makes them increase their own self-concept with respect to mathematics (p. 60). 60).

Part of the variety of possibilities of gamification focused on education is the use of digital tools. These tools allow the implementation of activities, organization, publication of materials and communication between those involved, whether coordinators, teachers, students and parents (Reyes Jofré, 2018). Gamification focused on education aims to successfully improve the learning of students, whether in face-to-face or virtual scenarios; there are several alternatives to provide conditions conducive to meaningful and sustained learning over time, providing opportunities and personal growth.

Conclusions

In conclusion, this study focused on describing how teachers at an Eloy Alfaro Educational Institution use gamification to teach mathematics to children. We have identified the techniques they use and explored the benefits of this strategy in the teaching and learning process. Gamification, by incorporating game elements in education, seeks to make the learning experience more attractive and participatory, greatly improving the academic performance of students, creating a dynamic with the subject, awakening interest in the subject and raising the problem-solving capacity of the context in which they develop.

Our findings suggest that this strategy can improve school performance, as well as students' formative and personal development. In a world where mathematics teaching faces challenges such as lack of motivation and detachment from the subject, gamification emerges as a promising tool to transform the way mathematics is taught and learned, fostering a more dynamic and collaborative learning environment. It is necessary that teachers incorporate this strategy more frequently to meet the expectations of all those involved, including students, teachers and parents, i.e. the educational community supported in an interesting and coordinated initiative, which opens the way to future forms of learning.

To achieve success, the first step is to create the necessary conditions, these conditions go hand in hand with the skills of people with a purpose, that recognize their role in society and with support generate opportunities, is the mechanics of education, we must only be actors of change.

Bibliographic references

Ardila-Muñoz, J. And "Theoretical assumptions for the gamification of higher education." International Journal of Research in Education, vol. 12, 2019, pp. 7184, https://doi.org/10.11144/Javeriana.m12-24.stge

Arias Gonzáles, José Luis, and Mitsuo Covinos Gallardo. Research Design and Methodology. Primera, ENFOQUES CONSULTING EIRL, 2021, Electronic book available in: www.tesisconjosearias.com

Aristizábal Z, Jorge, et al. "The game as a didactic strategy to develop numerical thinking in the four basic operations." Sophia, vol. 12, 2016, pp. 117-25, https://www.redalyc.org/pdf/4137/413744648009.pdf

Cenedesi, Mario Angelo, and Silvia Elena Vouillat. Research Methodology: from topic to publication of data. 36th ed., vol. 17, 2024, https://doi.org/10.32813/2179-1120.2024.v17.n1.a976

Chaves Yuste, Beatriz. "Review of gamification experiences in foreign language teaching." 33, vol. 8, November 2019, pp. 422-30, https://doi.org/0000- 0002-4442-9138.

Contreras Espinosa, Ruth, and José Luis (editors) Eguia. Experiences of gamification in the classroom. 2017, http://incom.uab.cat

Culqui Tibán, Daniela Lissette. Gamification as a methodological strategy in academic reinforcement activities in the area of Mathematics. 2023. Pontificia Universidad Católica del Ecuador, Ambato, https://repositorio.pucesa.edu.ec/bitstream/123456789/4254/1/MIE%20Cul qui%20Ti ban%20Daniela%20Lissette.pdf

García Casaus, F., et al. "Gamification in the teaching-learning process: a theoretical approach". Dialnet Repository, July 2020, pp. 16-24, https://dialnet.unirioja.es/servlet/articulo?codigo=7643607.

Hernández Sampieri, Roberto, and Christian Mendoza Torres. Research Methodology, the quantitative, qualitative and mixed routes. 2018, https://blogs.ugto.mx/rea/wp-content/uploads/sites/71/2021/11/Cap-3-Sampieri018.pdf. P.E-919087654123.

Hernández-Peñaranda, J. O., et al. "Use and benefits of gamification in mathematics education." Eco Matemático, vol. 11, June 2020, pp. 30-38, https://doi.org/https://doi.org/10.22463/17948231.3200

Hurtado de Barrera, Jacqueline. Methodology of Holistic Research. 3rd ed., SYPAL, 2000.

Jinez, Fernando. "Uso de la Gamificación para fortalecer el aprendizaje de las Matemáticas en los estudiantes de 1RO BGU." Indoamerican Technological University, vol. 1, October 2023, pp. 77-81.

León Flores, Daniela Tahís. Proposal of gamification in a didactic unit in the subject of mathematics with students of 6 EGB. 2022. Pontificia Universidad Católica del Ecuador, Esmeraldas, https://repositorio.puce.edu.ec/server/api/core/bitstreams/4f9c40e8-a51a-4a81-9592- aba025771ba9/content

López López, Mónica Yazmín. "The importance of gamification as a teaching-learning technique at the higher level." Insigne visual, vol. 24, July 2019, pp.4957, http : //www.apps. buap.mx/oj s3/index.php/insigne/article/view/1442/1046

Polo Escobar, Benjamín Roldan, et al. Transversal competences in the university educational context: a critical thinking from the principles of gamification. July 2022, p. 178, https://doi.org/https://orcid.org/0000-0001-5056-9957

Ramírez, María del Rocío, and Héctor Olmos. Cognitive functions and motivation in mathematics learning. July 2020, pp. 51-63.

Reyes, David. "Gamification of virtual learning spaces." Virtual Training Center, 2018, UMCE; david.reyes_j@umce.cl - davide.reyesj @gmail.com.

Romero Aimacaña, Jhon, and Luis Velasco Bautista. Gamificación e Innovación Pedagógica. 2023. Technical University of Cotopaxi, https://repositorio.utc.edu.ec/bitstream/27000/11572/1/PP-000322.pdf

Vásquez Unda, Marco Miguel. Gamificación y estándares de aprendizaje del área de matemáticas en estudiantes, U.E. Veinticuatro de Mayo, Santo Domingo. Ecuador 2021. 2021. Universidad César Vallejo Lima - Perú, https://repositorio.ucv. edu.pe/bitstream/handle/20.500.12692/78247/Vasquez   UMM-SD.pdf?sequence=1&isAllowed=y

Printed by Books on Demand GmbH, Norderstedt / Germany